CAREER AS AN ENGINEER

MATERIALS AND METALLURGICAL

THROUGHOUT HUMAN HISTORY, CIVILIZATIONS have been built with the best materials available at the time. It started with a handful of natural substances – rocks, mud, wood, skins, and straw. Today, there is a vast range of materials available, from vulcanized rubber for our shoes to ceramic composites in military aircraft. Where do our modern materials come from? Some still come from the earth in the form of ores that are extracted, refined, and used to fabricate all kinds of products. The rest arise from the imagination of materials engineers.

Materials engineers are on the cutting edge of technology in virtually every field. They develop materials with precise combinations of mechanical, chemical, and electrical properties that make other advances possible. Metals, plastics, ceramics, nanomaterials, and semi-conductors are just a few materials that these professionals continue to develop and enhance.

Materials engineers are often the unsung heroes that make life better for countless millions of people. Here are some examples of what materials engineers have accomplished:

- Invented clothing that repels mosquitoes
- Created more flexible materials to enhance the maneuverability of skis and snowboards
- Developed nanosized polymers that help repair shattered bones
- Developed artificial skin grafts that help burn victims heal faster
- Created stretchable material for clothing that returns to its original shape

In addition to developing new types of materials, materials engineers look for new ways to use existing materials. In the manufacturing environment, they are often tasked with selecting the most appropriate materials to produce particular items, whether that is computer chips or golf clubs. In the mining industry, they work on more efficient and eco-friendly ways of extracting minerals and processing them. Many materials engineers are also involved in research, studying the properties and structures of substances in order to create needed materials that do not yet exist, or to determine the cause of material failure.

Materials engineers work in a variety of environments. Those focused on research and development work in university research laboratories or industry development labs. Those associated with manufacturing may work in factories or offices where they have access to advanced computers and design equipment. Some materials engineers spend much of their time out of doors at mining operations that may be quite remote. While the environment and hours that they work can vary substantially with their industry and job, most materials engineers work regular hours.

As you might imagine, it takes considerable knowledge to do the work of a materials engineer. All of the physical sciences come into play, along with advanced math like calculus. The minimum education required for an entry-level job is a bachelor's degree in materials or metallurgical engineering. The only possible alternative is a degree in a closely related engineering field, such as chemical or biomedical engineering. In most cases, advanced research positions require at least a master's or doctoral degree. Completing internships and cooperative engineering programs while in school are the best methods for gaining valuable experience and making contacts with potential future employers.

As technical as this work is, it is important to note that employers look for good communications skills. There is a ton of paperwork associated with this work, as well as oral presentations to clients, managers, and team members. Communications are often between the materials engineer and non-engineering professionals. These people do not necessarily need to know about the chemical properties of a polymer. They may be more interested in possible future applications and profit potential. It is up to the materials engineer to convey the information they want to know in language they can understand.

After four years of college, a materials engineering graduate can feel confident that the hard work was worth it. This is an extremely diverse profession, offering a wide variety of opportunities. There is a steady demand for materials engineers who can design new materials and find new uses for existing ones. Most newcomers to the field are recruited on campus and do not need to pound the pavement looking for their first job after graduation. They are paid well, too. The average graduate can expect to earn between $60,000 and $70,000 for their first few years on the job. With the right combination of location, industry, and years of experience, that income can climb to $150,000.

WHAT YOU CAN DO NOW

MAKE HIGH SCHOOL COUNT. GO FOR CHALLENGING math and science courses like calculus, physics, and chemistry as well as algebra, geometry, and trigonometry. English classes are equally important. You will need to be able to read complex material and you will be writing detailed reports constantly. Computer science is a must. Check the admissions requirements for engineering majors at the colleges of your choice to make sure you are covered. Get a head start on engineering requirements any way you can with classes like analytical geometry, engineering science, and design.

Sculpture and welding classes are good for learning the properties of metals. Creating anything out of metal, from bronze sculptures to metal jewelry, is valuable hands-on experience.

Good grades are a must, but they are not the only important thing. Admissions officers look for well-rounded individuals with lives outside the classroom. Your high school years should include activities of all kinds. Join an engineering club and seriously consider going to an engineering summer camp. There are also excellent opportunities through special academic programs, such as Vanderbilt University's summer program, Preparatory Academics for Vanderbilt Engineers (PAVE). This summer program for high school students provides practical experience in engineering, computer skills, problem solving, and technical writing.

Participate in science fairs and engineering competitions such as FIRST (For the Inspiration and Recognition of Science and Technology) or the ones offered by the Technology Student Association. These events are a fun way of learning engineering concepts.

Start reading professional publications like *Materials Today.*

HISTORY OF THE CAREER

IN MANY CASES, THE ONLY EVIDENCE of ancient cultures is the materials that survived. Archeologists have been able to characterize and distinguish between peoples based on the predominant materials found in each location. Stone Age cultures were especially limited by which rocks they could find locally. As civilizations grew, and trade routes were established, more materials were acquired by trading.

Some of the earliest tools were daggers fashioned from the thigh bones of animals. Depending on the size of the animal, the daggers were used to hunt, fish, and butcher game. These daggers evolved into wood and stone axes, then copper, bronze, and iron implements. Mineral rock deposits, known as ores, became valuable. People who had access to ore and knew how to extract the minerals became powerful. This included early alchemists, who were tasked with finding ores and, more importantly, developing methods for metal extraction.

Metals and metalworking had been important to the building of the Roman Empire since the Bronze Age. Italy itself was not rich in metal ores, but as the Empire expanded its influence, it was able to acquire areas rich in minerals. One of the first descriptions of a metallurgical process was written by Roman goldsmith Benvenuto Cellini, who was known for turning gold into fanciful objects for dukes and popes. The most important Roman source of information is the *Naturalis Historia* of Pliny the Elder. Several books within this encyclopedia cover metals and metal ores, their occurrence, importance and development.

Metals of all kinds continued to be mined throughout the centuries, but the most significant developments in metallurgy focused on applications for iron. Of all the minerals discovered throughout history, iron has been the most important to the evolution of civilization.

The first systematic book on metallurgy, *Pirotechnia*, was published in the 16th century by Italian metallurgist, Vannoccio Biringuccio. However, it was German physicist, Georg Agricola, who became known as the "father of metallurgy" after his book *De Re Metallica* was published. He described the highly developed and complex processes of mining metal ores and metal extraction techniques of the time. Later, in the 17th century, Galileo's *Two New Sciences* talked about the strength of materials and kinematics. It was the first quantitative statements in the science of materials.

During the 18th and 19th centuries, metallurgists began to better understand the properties of metals. It was then that metallurgy as a science began. Multiple patents were filed, including a copper/zinc acid battery, hydraulic cement (stucco), and a process for producing metallic zinc by distillation from calamine and charcoal. Other inventions included the thermocouple, Portland cement, metallic aluminum, vulcanized rubber, silver-based photographic processes, and the first solar cells (using selenium waffles).

When metallurgists began to study alloys in the late 1800s, physical metallurgy evolved into a modern science. By the 20th century, huge advances were underway, such as the development of stainless steel, the discovery of a strong and lightweight aluminum, and new uses for magnesium. Great strides were also made in nonmetallic materials like ceramics, glass, plastics, and semiconductors. At this point, the field known as metallurgy became known as materials science because the same professionals were dealing with both metallic and nonmetallic substances.

Most engineering schools in the early 20th century had a department of metallurgy and sometimes also ceramics. The curriculum was primarily focused on the processes that underlie steel production. There simply was not enough scientific knowledge about other materials to include them as academic subjects. That changed after World War II when the study of polymers advanced quickly. Engineering schools toyed with the idea of creating new polymer science departments, but ultimately decided it was more sensible to support materials science as a new interdisciplinary field in its own right. In 1955, Northwestern University established the first materials science department intended to consider all materials of engineering importance from a unified point of view.

The field got a major boost in 1958 when President Dwight D. Eisenhower created the Advanced Research Project Agency (ARPA). The purpose of ARPA was to encourage the establishment of interdisciplinary laboratories (IDLs) on university campuses that would research materials and teach students how to conduct materials science research. Contracts to support the effort were awarded to 13 universities, and the program is considered a milestone in the history of research and development of materials science in the US.

Many innovations have emerged over the centuries, but one of the most interesting is powder metallurgy. It is a process in which metals are turned into powders, compressed, and then heat treated to fabricate a particular product. In the 20th century, the method led to the development of many new alloys and composite materials. The interesting thing about powder metallurgy is that it has been around for millennia. Although it did not take root as an industry until the mid-20th century, the Egyptians utilized iron powders for producing objects as early as 3000 BC, and the Incas of Equador created small objects of platinum with powder metallurgy methods.

Today's materials and metallurgical engineers are concerned with ways to conserve our dwindling natural resources and protect our environment. Many of the once mineral-rich underground deposits have already been depleted, but bridges, buildings, vehicles, tools, and machines are still made with metals. Engineers are focused on finding more efficient, automated processes of metal recovery and reuse. It is the most complex kind of recycling. In addition, they are concerned with environmental protection and are looking for ways to reduce the pollution created by the traditional extraction processes.

WHERE YOU WILL WORK

WHERE DO THESE ENGINEERS WORK? IN just about every manufacturing industry throughout the country. That is because one out of four jobs in the US involves the manufacture or use of metal. Opportunities for these professionals can be found in aerospace, automotive, heavy equipment, electronics, metals casting heat treatment, and metals production companies. Not all materials and metallurgical engineers work in manufacturing. Many work in research institutes, universities, engineering consulting firms, and various agencies of the federal government.

The work environment varies widely depending on the

type of work and the industry. Some work in offices, using cutting edge computer and design equipment. Those working in physical metallurgy are typically found in labs and manufacturing plants, conducting experiments and studies on extracted metals. Extractive metallurgical engineers are usually located in steel mills, refineries, ore treatment plants, or smelter plants. In some cases they work at remote mining sites. Process engineers work anywhere metal is used, from welding shops to rolling mills.

Some materials engineers dedicate their work to research and development. These specialists spend much of their time in labs and libraries. Those who go into academia may spend time in classrooms, but it is common for them to work as part-time consultants on the side.

Materials engineers usually work full time, on a regular 40-hour per week schedule. In some positions, such as industrial refining, night shift work may be required. The workday may occasionally extend into the evening or weekend when there are deadlines to meet, or if special experiments require more attention.

THE WORK YOU WILL DO

THE WORK OF MATERIALS ENGINEERS involves raw materials that are used to create products of all kinds, such as snow skis, aircraft wings, biomedical devices, golf clubs, cars, and electronic devices. They work with all kinds of existing materials like metals, ceramics, plastics, composites, semiconductors, and nanomaterials (extremely small substances). The most basic tasks include finding sources of materials (like mineral deposits), selecting the most appropriate material for specific products, and developing new ways to use existing materials. For example, they may find ways to improve the strength and safety of metals that are used in healthcare, transportation, and defense.

Materials engineers also study the properties and structures of materials in order to create new materials that are needed to solve problems in several different engineering fields, such as mechanical, chemical, electrical, civil, nuclear, and aerospace. Using computers, they can study and create materials at an atomic level using an advanced process to model and replicate characteristics and components. Testing is an important part of this process. Specialized software is used to run simulations and conduct tests on raw materials or finished products to ensure their quality and safety.

Computers are essential tools for materials engineers. In addition to creating and testing new materials, they are used to generate part specifications, monitor performance, control efficiency, and create and analyze designs.

Materials engineers are closely aligned with business. Their job is to develop economical solutions to technical problems, and to apply scientific principles to commercial applications to meet consumer needs. In other words, they help develop new products for the marketplace. They consider several factors during the process, including precise functional requirements, design, and the overall cost, safety, reliability and effectiveness of the product.

Beyond development and design, materials engineers work in production, quality control, and maintenance. They supervise production in factories, test for product quality, and determine the causes of failing products.

Other common job responsibilities include:

- Prepare proposals, write reports, and perform other managerial tasks

- Estimate cost and time for projects

- Review new product plans and make recommendations for material selection based on strength, weight, heat resistance, electrical conductivity, and cost

- Monitor how materials perform and evaluate how fast and in what way they deteriorate

- Determine causes of product failure and develop ways to solve the problem

- Assess how certain materials perform in a range of conditions

- Evaluate the impact of materials processing on the environment

Specializations

Materials engineering is a vast field. It is impossible to be an expert on every kind of material and all its properties and uses. Therefore, materials engineers usually specialize in a particular material, such as graphite, metal and metal alloys, ceramics and glass, plastics and polymers, and naturally occurring materials. For example, those working with ceramics might help to create products such as fiber-optic communication lines or high-temperature rocket nozzles. By contract, engineers with expertise in semiconductor processing help develop new microelectronic materials for computing, sensing, and related applications.

Metallurgical engineers specialize in metals, such as steel and aluminum, usually in alloyed form with additions of other elements to make them stable and strong.

Metals of many varieties are an important part of most modern manufacturing processes, although metallurgical work is slightly different from ordinary metalworking. Metalworking is the process manufacturers use to shape and craft metal into the shapes most people recognize and use every day, but metallurgy is the science behind the metalworking. It involves taking a particular metal and getting it into the right chemical and alloyed state so that it can be manipulated and used for specific purposes. This requires scientific knowledge and hands-on testing. Because metal is the primary material in so many different kinds of products, metallurgical engineers often subspecialize in a particular type, such as automotive or sporting goods.

The two main types of metallurgy are physical and extractive. The former concentrates on the physical properties and structures of alloys and metals. The latter is devoted to separating metal from ore. Metallurgical engineers are rarely involved in both types.

Physical metallurgical engineers (also known as physical metallurgists) are the pioneers of metallurgy, always finding ways to create something new.

Most metals cannot be utilized in their natural state. They need to be blended with other metals and nonmetal substances to create alloys. The best alloys are strong yet lightweight, in order to be useful for most products whether they be airplane wings or smart phones. Physical metallurgists get to work after metals have been extracted and refined. They spend their days in labs, working to develop new alloys or to improve production processes for welding, melting, and casting. Physical metallurgists also develop new applications for metals.

In the lab, physical metallurgists use scientific methods to study the relationship between the physical structure and other characteristics of metals to determine whether they can be used in thermal, electrical, or magnetic applications. They conduct X-ray and microscopic experiments to test the components for impurities and defects, and to determine how long the metal could be expected to last when used in particular products. The studies are also used to devise ways to transform existing alloys to give them certain desired characteristics that they may not already possess, such as corrosion resistance, malleability, hardness, or durability.

Extractive metallurgical engineers (also known as chemical metallurgists or extractive metallurgists) work with natural metals.

Their workplace is usually in a steel mill, ore treatment plant, or refinery. Their primary job is to create and oversee the extraction processes used to separate metals from ores. Every ore is different and it is up to the extractive metallurgist to determine which method will produce the greatest concentration of ore by separating minerals from dirt, rock, and other unwanted materials. They work closely with mining engineers, often supervising and controlling the separation and final purification processes in commercial mining operations. Their work often takes them to remote sites and plants.

It is common for extractive metallurgists to be involved in the design of treatment plants and refineries. They may select the machinery to be used and oversee its installation, train refinery workers, and supervise the processing operations. They monitor operations and employ new methods or modifications to improve efficiency if necessary.

Some extractive metallurgists do work in labs using

advanced equipment like spectrographs and electron microscopes to research ways to use new sources of metals, such as reclaiming magnesium from seawater. They also research better ways to separate metals by processing small batches of ores in the lab and using computer software to analyze the efficiency of various methods.

A growing responsibility for extractive metallurgists is to find better ways to reclaim materials from solid wastes for recycling. After thousands of years of exploitation, high-grade ores are becoming depleted. It is the extractive metallurgist's mission to find new ways to extract metals from low-grade ores and to recycle metals that are considered scrap material. These professionals have also made great strides over the past two decades in improving the refining processes so there is much less environmental damage from waste materials.

Materials engineers can specialize in other types of materials as well. The most common are glass, composites, and polymers.

Glass materials engineers focus on the use of glass, ceramics, and all its related composites.

They are generally involved in research and development or in manufacturing. Those in research and development develop and test new materials that can be used in place of traditional glass in applications where enhanced performance or strength may be required. For example, glass engineers developed the specialized glass used to make LCD flat-panel displays. In the process, they might create models or test structures and conduct experiments to ascertain if the material can withstand certain types of forces, temperatures, or other unique factors. A glass engineer may also develop new equipment that can be used to improve the glass or ceramic production process.

Glass engineers working in manufacturing typically design products and oversee the production of structures or products that utilize glass or ceramics. For example, buildings with large glass surfaces or exteriors that involve glass require substantial input from a glass engineer. It would be the engineer's job to ensure the glass would connect with other materials in a way that keeps water and air out, yet strong and durable enough to handle normal building forces. Glass engineers work on more than skyscrapers. Glass and ceramics are used in an endless variety of products, such as fiber optics, electronics, eyeglasses, vehicles, and spacecraft.

Experienced glass engineers are often called upon to provide training on the factory floor or in another workplace environment.

Composite engineers study materials (commonly known as composites) made from two or more materials with different properties that, when combined, produce a single new material with structural characteristics that are different than either of the original components.

Concrete is a composite, made of loose stones and cement. It is inexpensive and will not compress or shatter, but if stretched will quickly break apart. Steel is expensive, but highly resistant to stretching. When added to the concrete, the result is an economical material that can be safely used in large structures.

The composite engineer's job is to develop materials with specific characteristics that suit the end product. For example, in the construction industry one of the most common composites is reinforced concrete made of loose stones, cement, and steel bars. It is an economical material that is safe to use for large structures because it will not compress, shatter, or break apart under tensile load. An engineer in the aerospace industry, on the other hand, would create advanced composites made of various chemicals for use in wind turbine blades and in spacecraft that need to be able to withstand extraordinarily demanding environments. Other composites have different characteristics that make them suitable for things like boat hulls, performance bicycle parts, imitation skin, prosthetics, or imitation granite countertops.

Composite engineers play key roles in facilities that produce composites and any factories that use composites to fabricate products. In either case, they usually supervise the production process and determine which materials and processes should be used. A wide range of materials is used to make composites, but generally include things like carbon, resins, and fibre-based materials. Each component will affect the characteristics of the composite material and the end product. The composite engineer is responsible for choosing the most appropriate base resins, additives, and reinforcing materials that will produce finished materials with the required characteristics. The engineer will also determine the best manufacturing process for the particular project.

Polymer engineers study materials made up of many repeating molecules strung together.

The word polymer, which has Greek origins, means "many parts." There are natural polymers like muscle fibers, cellulose, shellac, wool, and DNA. However, in the materials engineering world, they are synthetic materials like plastics, rubber, carbon fiber, and nylon.

Polymer engineers design new polymer materials and resolve problems related to existing polymer compounds. What makes their job fascinating is how differently polymers act depending on the kinds of molecules that are put together. They may be rubbery, sticky, soft, hard, tough, stretchy, or any combination of characteristics.

Change the combination and you will get silly putty instead of a soft, fuzzy sweater.

Polymer engineers often specialize either in polymers classified as thermoplastics or thermosets. Engineers who specialize in thermosets create materials used for building like fiberglass and aramids. These kinds of materials need to be lightweight, but strong enough to bear a high weight load. Engineers who work with thermoplastics create materials that are often used in consumer products. Thermoplastics are a type of transparent, low-density polymer. A common product made of this material is food storage containers. There is, however, a vast array of polymer materials and job roles vary widely by the type of polymers and the industry.

STORIES OF WORKING ENGINEERS

I Work in the Aerospace Industry

"I focus on research and development, specializing in the processing of high temperature ceramics like carbides, nitrides, and borides. These are the kinds of materials that work well in thermal protection systems for hypersonic flight and re-entry space vehicles. The work is fun because I get to experiment with advanced techniques like spark plasma sintering. That's running tens of thousands of amps through graphite die to create dense ceramic parts. I also design and build lots of specialized equipment.

R&D is not just about research. It involves a lot of hands-on work, too. I'm usually in the machine shop at least a couple of days a week, cutting and machining the parts for testing. Sometimes I use plasma wind tunnels to test parts for flight properties. In addition to the exciting hands-on stuff, there is a lot of writing and making presentations. Basically, I make stuff, break stuff, burn stuff, and write about it.

My advice to aspiring materials engineers is to apply for internships everywhere possible. Don't restrict yourself to local companies. Big companies often pay room, board, and travel expenses for their out-of-town interns. As an intern, your job is to learn as much as possible. Pay special attention to how tools work, inspection procedures, safety protocols, testing methods, and processes. If you don't understand something, ask.

Don't worry if you don't get into your desired specialty right away. Start where you can and then transfer when you can. No one will question that. It's done all the time. I spent two years in medical devices before landing this job, but transferred as soon as possible. It will be harder to transition between industries later in your career."

I Am a Manufacturing Manager

"My company produces materials for ballistics armor used by the military and for private companies in the armor business. My responsibility is to create and implement production plans and to ensure compliance with ISO (Industrial Organization for Standardization).

I started out in college wanting to be a chemical engineer. I wanted to develop new materials and dreamed of helping build faster race cars. I didn't realize chemical engineers don't do that and I didn't know materials engineering even existed. One of my professors set me straight and I switched majors. I fell in love with the field during the two internships I did. They were completely different industries and I learned that you could actually change materials to suit different applications by making small modifications to the composition. That to me is fascinating.

After 12 years in the business, I am still fascinated with the work. Materials engineering is always at the forefront of technology and that means it's always changing and evolving. In the 1990s, the excitement was about nanotechnologies. Today, it is more about products like fuel cells. Discovering new technologies and related materials is fun, but the real challenge for materials engineers is to figure out how to apply the technology to improve the world we live in.

I wish someone had told me that studying engineering was going to be harder than I expected. I was always a good student, but college was really hard. I would advise anyone thinking of this profession to get well prepared in high school. Take AP courses, get as much advanced math and science as you can, and don't dismiss the importance of English classes.

Materials engineering is very rewarding and there are lots of options, too. You can choose to work in any field that interests you from sports to medicine to aerospace (and yes, racing cars). You don't have to do engineering forever. For example, you could get into marketing and introduce a new medical device or some exotic material to manufacturers."

I Am a Metallurgical Consultant

"I am a materials engineer with an advanced degree in metallurgy. I like to figure out problems and I don't like to stay in one place for too long. That's why I started my own consulting business, doing research and troubleshooting for international clients. Because I work for a variety of clients, there is no typical day. However, most of my projects involve failure analyses. I am called in when a mechanism has failed and find ways to fix the problem. I do this by examining materials samples, slicing them down, and using scanning electron microscopy to determine the cause of failure. It is a lot like detective work, which makes it perfect for a technically inclined person with a strong sense of curiosity.

I much prefer consulting to being a full-time employee. The money is much better, and since I'm in control of my schedule, I can make sure my hours are short and the days are busy. I also enjoy working with the great people I meet. I've worked in nine different countries so far and I've learned a ton from every experience. An added bonus is the nature of the clients, who are generally very high caliber people. That means I travel first class and stay in the best hotels – all paid for by the client.

There are a lot of options for young materials engineers today. You don't have to know exactly what kind of work you want to do starting out, but it helps. The sooner you know what your goal is, the sooner you will achieve it. The best way to figure it out is to intern in as many places as possible. Don't expect to like every position. Sometimes it's learning what you don't want to do that sets you on the path to your ideal career choice."

PERSONAL QUALIFICATIONS

SUCCESSFUL MATERIALS AND METALLURGICAL engineers have a high level of intellectual curiosity and analytical minds. In general, they are interested in nature and the physical sciences, and they possess a technical aptitude. They are also patient – which is important since much of their work involves long-term projects and studies. People who are happy in this career have other personality traits in common, including the following.

People obsessed with math and science can do well in this field. Math is an integral part of the work. The principles of calculus and other advanced topics in math are used daily for analysis, design, and troubleshooting. The most successful materials engineers are curious and love to solve puzzles. Solving engineering problems is not always easy because there is often more than one right answer. A good head for math and science is certainly necessary, but that knowledge is worthless if you do not enjoy the challenge of applying it in the context of real world problems.

Material and metallurgical engineers are creative and analytical. Both traits are used to determine how materials can be used under a variety of conditions. The work often involves trouble shooting. Analytical skills are particularly important for figuring out why a product failed, coming up with a new design, and patiently conducting exhaustive tests to make sure failure does not occur again.

Good communications skills are very important, both written and oral. Materials engineers spend a good deal of time writing detailed plans and reports. Some will be read by other engineers, technicians, and technologists, but that is not always the case. It is best to avoid engineering lingo and explain technical concepts in a clear way that anyone can understand. This is also true when speaking. Materials engineers must be able to state directions clearly and discuss progress with managers who may not have an engineering background.

It is a common misconception that engineers work with things, not people. The fact is material and metallurgical engineers must be comfortable working both alone as well as on a team. The stronger your people skills, the easier this will be. Meetings and planning sessions occur on a daily basis, sometimes several times a day. Many times you will be interacting with a wide range of professionals outside of engineering, from mining foremen to plant supervisors. If you work as a consultant, you will need to hone your interpersonal skills to meet with clients, initially selling them on your services, and then periodically making progress reports on what you are accomplishing. You will have to be flexible and able to look at things from the client's point of view.

This is a career for lifelong learners. Keeping up with the latest research developments and laboratory techniques is part of the job. The best materials and metallurgical engineers are avid readers, dedicated to staying abreast of industry advances.

ATTRACTIVE FEATURES

TWO BIG BENEFITS OF THIS CAREER are the good job prospects and the earnings. Most materials engineering graduates do not have to go looking for jobs because they are typically recruited right on campus, usually before they graduate. Preparing for this career requires a rigorous education so it is good to know there will be a job waiting. This is a stable field that is not affected much by seasonal fluctuations or economic downturns. The money will be good right from the start. The average starting pay for new materials engineers is in the $60,000 to $70,000 range. That is considerably higher than you can expect from most careers right out of college. Earnings will increase over time as you gain experience.

One of the amazing things about this career is the opportunity to work in a wide variety of industries – just about anything from environmental studies to sporting goods. Materials of all kinds are used to create everything we touch, and many things we do not even see. Materials engineers can therefore work in companies that develop ordinary things like shoes, chairs, eyeglasses, or cars, or they can get involved in things that are on the cusp of discovery. Just as the field is diverse, so are the kinds of people you work with. Materials engineers could be working side by side with other engineers – or they could be collaborating with fashion designers, medical scientists, professional athletes, or astronauts.

This is meaningful work. Materials engineers make important contributions that often help improve the quality of life for millions of people. Their research has led to improved safety of cars and airplanes, the invention of new life-saving medical devices, more efficient recycling techniques to preserve our resources, and the ability to reach other planets.

UNATTRACTIVE ASPECTS

THIS CAREER HAS MUCH TO OFFER with few drawbacks. Some consultants complain that doing project-based work means frequent relocation, but they are compensated with higher pay than their salaried counterparts get, which is the main draw of consulting in the first place. Others worry that since many jobs are associated with manufacturing, which is a sector in decline, they may have difficulty finding jobs in the future. However, this is an ever-expanding field with myriad opportunities for those who are qualified.

This is a broad field loaded with possibilities. You can choose from a wide range of industries – aerospace, software, entertainment, medical devices, and sportswear. The only problem is there is no way someone can be an expert in everything. While all materials engineers are well trained in the basics, the specific experience required for a particular role can be very different. That often means additional education, training, or experience. Plus, all materials engineers need to keep up with advances in technology through continuing education.

EDUCATION AND TRAINING

THE EDUCATIONAL REQUIREMENTS for this career are very clear cut. Materials engineers must have a bachelor's degree in materials or metallurgical engineering or a closely related engineering field. There are no alternative pathways. ABET-accredited programs are offered by more than 80 universities and colleges in the US. There is a variety of programs, with emphasis placed on different areas within the field. For example, some programs prepare students for practical design and production work, while others focus on theoretical science and math that might be used in research. It is therefore advisable to explore as many programs as possible to ensure a good fit for your career ambitions.

All programs include classroom and laboratory work building on engineering principles. The first two years of college set the groundwork for analytical thinking, with courses such as chemistry, physics, and mathematics. Introductory engineering classes are also part of the first year curriculum. From the sophomore year onward, about one out of three courses will be focused on materials and metallurgy. During the next two years, courses become more highly concentrated in materials and metallurgical engineering subjects. There will be more design courses that incorporate classroom study and computer work, and more hands-on lab classes, too. Lab work will focus on applying the principles learned in the classroom to practical design and production.

More than half of new materials and metallurgical engineers enter the field with a bachelor's degree. However, a master's degree is becoming increasingly important, and many graduates choose to continue on for an advanced degree. It generally takes two additional years of study to earn a master's degree. Some schools offer a combined five-year program that grants both a bachelor's and master's degree to graduates.

For those who aspire to a faculty position in a college or university or want to do research and development, a doctoral degree is needed. Earning a doctoral degree requires an additional four years of study and research beyond the bachelor's degree.

Most schools offer internships and cooperative programs in partnership with industry. Students in these programs get valuable practical experience while completing their education. Participation can also be very helpful for getting entry- level positions.

Certification, Registration, and Licensing

Certification is recommended, though not required to work as a materials and metallurgical engineer. Employers like to see it because it provides solid evidence of an individual's abilities. Certification in this field is available through ASM International and other materials science organizations. The Society of Manufacturing Engineers offers two designations to metallurgical engineers who work in manufacturing and who meet education and experience requirements: Certified Manufacturing Engineer and Certified Engineer Manager.

All 50 states and the District of Columbia require engineers to register if their work may affect the life, health, or safety of the public. Requirements for registration include a degree from an accredited engineering program, four years of relevant work experience, and a passing score on a written exam conducted by the state.

A license is not required for entry-level positions, nor is it very common among materials engineers. However, for those who aspire to higher management positions or want to work independently, a Professional Engineering (PE) license can be acquired after several years in the field. Those who do pursue this license are thereafter known as Professional Engineers (PEs) and are qualified to supervise other engineers, sign off on projects, and provide services directly to the public. Each state issues its own licenses and while requirements may vary somewhat, they generally include successful passage of two exams, a degree from an ABET-accredited engineering program, and relevant work experience (typically at least four years). Some states also require continuing education for engineers to keep their licenses.

EARNINGS

THE AVERAGE YEARLY SALARY FOR materials engineers in the US is around $95,000. Individual salaries are mainly influenced by experience level. For those starting out, in the first five years the median annual salary is about $65,000. The range is typically $60,000 to $70,000 for new graduates.

At the other end of the spectrum, the most experienced engineers often earn more than $150,000. That is for late career engineers, usually with more than 20 years on the job. In between is where most materials engineers fall. Someone with five to 10 years of experience will earn an average of $85,000. That number rises about $15,000 over the following 10 years.

Total cash earnings are often boosted beyond the base salary. Bonuses are common and often approach

$10,000. Profit sharing is also common with proceeds averaging around $15,000. Both may be tied to individual performance so the most successful employees reap the biggest benefits.

Geographic location also has a large influence on pay levels. The top five highest paying states are California, Louisiana, Maryland, Colorado, and the District of Columbia. The average pay rate in any of these states is more than $115,000 a year. New York State is home to the smallest salaries in the field, lagging the national average by almost 20 percent. The particular city also matters. One of the easiest ways to increase your salary as a materials engineer is to move to Denver or San Diego.

Other factors that affect earnings is the industry and the size of the company. For example, a job in primary metal manufacturing would pay around $75,000, while the same job in transportation equipment manufacturing would pay roughly $100,000. The best paying jobs are in computer and electronic product manufacturing. Consulting services and research and development fall somewhere in between. As a general rule, bigger companies in the Fortune 500, like Boeing and General Electric, tend to pay the best.

Materials and metallurgical engineers enjoy generous benefits. Medical coverage, including dental insurance, is always part of the package. Paid vacations and holidays, sick days, and retirement plans are also common. Some companies offer tuition assistance for continuing education. Others pay for membership and expenses for participation in professional associations.

OPPORTUNITIES

THIS IS AN EXTREMELY DIVERSE PROFESSION, offering a wide variety of opportunities for careerists with an interest in science, technology, and engineering. Materials engineers will continue to be needed to design new materials for use both in traditional industries like manufacturing and aerospace, and in those focused on new medical or scientific products.

Metallurgical engineers are particularly well positioned. It has long been recognized that modern societies need an abundant supply of every imaginable type of metal and alloy. Therefore, jobs for metallurgical engineers will be found in every enterprise that produces, buys, sells, refines, or manufactures metals or metallic products. Because metals are used in so many industries, employment can be in large companies or small consulting firms.

The aerospace industry has long been a rich hunting ground for job seeking materials engineers. It started with a focus on metallurgy, but has expanded to encompass other materials, such as biomaterials and electronic materials. The demand is still primarily in research and development, though engineers who specialize in process refinement, materials testing, and characterization are sought after. That demand is expected to grow for decades to come. Opportunities in the aerospace industry can be found in federal agencies like the DOD, FAA, and NASA. In the private sector, most jobs are located in the New England area, which is home to major manufacturers of helicopters, jet engines, aerial drones, missiles, spacesuits, and life support equipment systems for the space station. These technologies require complex materials processing techniques and the ability to select appropriate materials for unconventional applications.

Materials engineers are increasingly needed for research and development in cutting- edge fields like biotechnology, three-dimensional printing, and nanotechnology. Medical implants are a particularly interesting area that requires the expertise of materials engineers to develop new materials. There is also a lot of demand in the micro/optoelectronics industry, which includes things like semiconductors, sensors, and other functional materials for electronic applications. Companies like Varian, IBM, Fisher Scientific, and many others are always recruiting materials engineers. Another small, evolving area that needs materials engineers is the fuel cell industry.

There is a finite amount of metal on the planet. In the past, only higher-grade ores have been used because they are the most profitable. As the world's ore deposits become further depleted, engineers are increasingly needed to develop new methods of processing lower--grade ores. They will be needed to locate new sites and devise new alloy combinations. Some will find jobs developing new methods of recycling scrap metals, designing nonpolluting processing systems, and developing better cleanup methods for existing plants.

An accredited college degree program in materials engineering provides the foundation for a prosperous career. Those who are trained in computer modeling should experience the best employment opportunities. Computer modeling and simulations are increasingly used in place of extensive and costly laboratory testing. Job prospects are also best for those who have participated in internship programs while in college. Employers will always choose the applicant with experience over the one who has yet to apply their classroom instruction to real projects.

GETTING STARTED

MOST MATERIALS AND METALLURGICAL ENGINEERS find their first job through their college job placement office. Check in regularly for newly posted job openings. Many students are also recruited on campus. Recruiters usually make regularly scheduled visits. Check with your advisor or the job placement office to learn when recruiters are expected, then sign up for interviews. Practice your interviewing skills and ask your advisor for feedback to help you polish your presentation skills. Remember, good communications skills are highly valued in the engineering world.

Most colleges sponsor job fairs at least twice a year. Make sure you attend all of them starting in your sophomore year. To stand out among the many students who will also be participating, get involved in projects outside the classroom. For example, join engineering organizations that sponsor special events or competitions. If the events coincide with the type of companies you want to work for, so much the better. That will provide relevant experience to add to your résumé. Prepare a 30-minute speech about yourself and the work you have done.

Make contacts and get experience through internships. Start applying for internships right away. Although it is not easy, it is possible to get internship offers as a sophomore or even a freshman. The key to finding one is by networking. If you have been doing all you can to prepare for this career, you already have plenty of contacts. Start with your professors and get in touch with every instructor and supervisor you ever had in summer camp or in other engineering events. Cultivate more contacts by attending professional association events and meetings. If you cannot get an internship, look for short-term jobs, freelance assignments, or any kind of side projects that will provide experience.

Some materials and metallurgical engineers find jobs the old-fashioned way, by taking the initiative and applying directly to companies. Look for job listings on professional association sites and job boards. There will be plenty of posts on the general sites like Indeed, but you will save time by sticking to job sites that cater to engineers. There are also professional recruitment agencies that specialize in engineering jobs.

Direct applications can get lost among the inflow of résumés every human resources department receives daily. To stand out, build a career portfolio. A career portfolio includes the usual résumé items like a bio page, recommendations, and information about your education, internships, volunteer work, language proficiency, and other things that make you a viable candidate. A portfolio goes well beyond a résumé, providing actual evidence of your technical and non-technical skills. Include examples of projects you have worked on with flowcharts, tables, videos, and links to your GitHub profile. Instead of simply listing your internships, include photos, links, and an essay that describes how your work benefitted the business. Discuss why you chose the internship, how you prepared, what challenges you encountered, and how you overcame them. Park your portfolio online, so you can include it on your LinkedIn profile or provide the link upon demand.

ASSOCIATIONS

■ **Accreditation Board for Engineering and Technology**
http://www.abet.org

■ **American Institute of Mining, Metallurgical, and Petroleum Engineers**
www.aimehq.org

■ **ASM International**
https://www.asminternational.org

■ **Materials Research Society**
http://www.mrs.org/home

■ **The Minerals, Metals and Materials Society**
http://www.tms.org

■ **Society of Manufacturing Engineers**
http://www.sme.org

■ **Society for Mining, Metallurgy, and Exploration**
http://www.smenet.org

■ **Technology Student Association**
http://www.tsaweb.org

PERIODICAL

■ **Materials Today**
https://www.materialstoday.com

WEBSITES

■ **FIRST (For the Inspiration and Recognition of Science and Technology)**
https://www.firstinspires.org

■ **GitHub**
https://github.com

■ **Industrial Organization for Standardization**
https://www.iso.org/home.html

■ **Preparatory Academics for Vanderbilt Engineers (PAVE)**
https://pave.vanderbilt.edu/ayindex.php

www.ingramcontent.com/pod-product-compliance
Lightning Source LLC
Chambersburg PA
CBHW071200220526
45468CB00003B/1097